READY SET STAAR

FOR TEXAS
SCIENCE SUCCESS

NATIONAL
GEOGRAPHIC

School Publishing

PROGRAM CONSULTANTS

Randy Bell, Ph.D.

Kathy Cabe Trundle, Ph.D.

Judith S. Lederman, Ph.D.

David W. Moore, Ph.D.

Grade 3

REPORTING CATEGORY 1
MATTER AND ENERGY

PAGE 4

PAGE 10

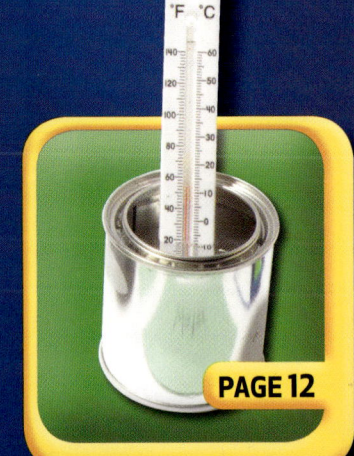

PAGE 12

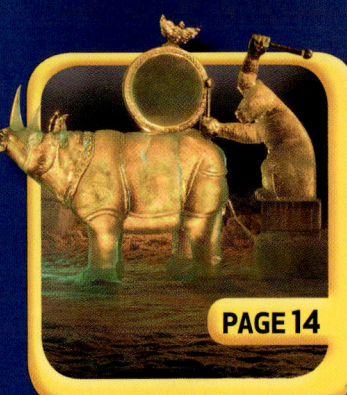

PAGE 14

Grade **3**

REPORTING CATEGORY 2
FORCE, *MOTION,* AND ENERGY

PAGE 30

PAGE 32

PAGE 34

PAGE 38

Grade 3

REPORTING CATEGORY 4
ORGANISMS AND ENVIRONMENTS

PAGE 64

PAGE 68

What are some solids, liquids, and gases in this photo of the Golden Gate Bridge in California? The concrete at the base of the bridge and in the roadway is a solid. The steel in the towers, the frame of the roadway, and the cables are also solids. The water in the ocean and in the tiny droplets in the fog are liquids. The air is made of gases, but they are invisible.

MATTER AND ENERGY

REPORTING CATEGORY 1: MATTER AND ENERGY

The student will demonstrate an understanding of the properties of matter and energy and their interactions.

3.5 MATTER and ENERGY

The student knows that matter has measurable physical properties and those properties determine how matter is classified, changed, and used.

STATES OF MATTER

Take a look at the photographs. Everything you see in the photographs is made of **matter.** Matter comes in different forms, called states. The **states of matter** include solids, liquids, and gases.

The large rock on the beach is made of basalt. The rock is a solid. The water in the ocean is a liquid. The air all around us is made up of different gases.

SUPPORTING STANDARD TEKS 3.5.C:
Predict, observe, and record changes in the state of matter caused by heating or cooling.

READY SET STAAR

Each state of matter has unique properties. A solid is matter that has a definite shape. A liquid is matter that can flow and takes the shape of its container. A gas is matter that spreads out to fill a space.

SOLID

This rock is a solid. If you moved the rock from one place to another, it would stay the same shape and size.

LIQUID

The dish soap in the bottle takes the shape of the bottle. A liquid does not have a definite shape when it is not in a container.

GAS

The air inside the bubbles is made up of different gases. The gases inside the bubbles spread out to fill the space inside the bubbles.

My science notebook

WRAP IT UP !

1. **Identify** What are three states of matter?

2. **Contrast** What is the difference between a solid and a liquid?

3. **Apply** You observe a block of ice. What state of matter is the ice? Explain.

CHANGES IN MATTER:
HEATING

Imagine you have a cup of ice. You set it out in sunlight on a warm day. As the ice is heated, it melts, or changes from a solid to a liquid. The liquid takes the shape of the cup.

Heating causes other changes in matter, too. When liquid water is heated, **evaporation** can happen. During evaporation, the liquid water changes into a gas called water vapor. Water vapor is invisible.

When the ice is heated, it melts. The ice changes state from a solid to a liquid.

voCAB

evaporation
(i-vap-uh-RĀ-shun)

Evaporation is the change from a liquid to a gas.

As the puddle is heated, some of the liquid water evaporates. The liquid water changes to water vapor, a gas.

SUPPORTING STANDARD TEKS 3.5.C:
Predict, observe, and record changes in the
state of matter caused by heating or cooling.

READY
SET
STAAR

Science in a Snap!

Melting

Make a small container out of foil. Place an ice cube in the container.

Predict what will happen to the ice cube. Record your prediction. After 1 hour, observe the ice cube.

Do your results support your prediction? How has the shape of the ice cube changed?

My science notebook

WRAP IT UP!

1. **Explain** Describe how matter changes state when a solid such as ice is heated.

2. **Explain** Describe how matter changes state when a liquid such as water is heated.

3. **Apply** An open container of water is left out on a table for a week. You observe that the water level has gone down. Explain what most likely happened.

INVESTIGATE
EVAPORATION

 **What happens to water as it is heated?**

Water can be a solid, a liquid, or a gas. It changes between these states of matter when it is heated or cooled. In this investigation, you can observe how heating affects liquid water.

MATERIALS

2 cups with water

plastic wrap

rubber band

marker

SUPPORTING STANDARD TEKS 3.5.C: Predict, observe, and record changes in the state of matter caused by heating or cooling.

READY SET STAAR

1

Draw a line on each cup to show how much water there is.

2

My science notebook

Put plastic wrap over 1 cup. Place a rubber band around the plastic wrap. Do not put anything over the other cup. Predict what will happen to the water in each cup. Record your predictions in your science notebook.

3

Put both cups in a sunny place. Observe the cups each day for 3 days. Record your observations.

4

After 3 days, observe how much water is in each cup. Is the water in each cup still at the line marked in step 1? Record your observations.

My science notebook

WRAP IT UP !

1. **Compare** How did the amount of water in the cups differ after 3 days?

2. **Explain** How did the water in each cup change as it was heated?

CHANGES IN MATTER: COOLING

Look at the water drops on the outside of the cold glass. When water vapor in the air touches a cold glass, the water vapor is cooled. Cooling causes the water vapor to change into drops of liquid water. **Condensation** is the change from a gas to a liquid.

Liquid water drops formed on the outside of this glass. The drops formed when water vapor in the air was cooled.

voCAB

condensation
(kon-den-SĀ-shun)

Condensation is the change from a gas to a liquid.

SUPPORTING STANDARD TEKS 3.5.C:
Predict, observe, and record changes in the state of matter caused by heating or cooling.

Look at the polar bear in the photo. It is standing on ice, a solid form of water. Ice forms when liquid water is cooled until it freezes.

Science in a Snap!

Freezing

Observe the shape of the water in a cup. Use a piece of clay to make a container. Pour water into the container. Put the container in a freezer. Predict what will happen to the water.

Take the ice out of the container and put it in a cup. Observe the shape of the ice in the cup.

 **Did your results support your prediction? How is the ice different from the liquid water?**

 *My science notebook*

WRAP IT UP !

1. **Explain** Describe how matter changes state when a liquid, such as water, is cooled.

2. **Explain** Describe how matter changes state when a gas, such as water vapor, is cooled.

CONDENSATION

 **What happens to water vapor as it is cooled?**

Water can change state when it is heated or cooled. In this investigation, you can find out what happens to water vapor, a gas, when it is cooled.

MATERIALS

| 1 can | crushed ice | water | graduated cylinder | thermometer |

SUPPORTING STANDARD TEKS 3.5.C:
Predict, observe, and record changes in the
state of matter caused by heating or cooling.

1

Fill a can half full with crushed ice.

2

My science notebook

Use a graduated cylinder to measure and pour 80 mL of water into the can. Predict what will happen to the outside of the can. Record your prediction in your science notebook.

3

Place a thermometer into the can and measure the temperature. Record the temperature.

4

Observe the outside of the can for about 3 minutes. Measure the temperature. Record your observations.

My science notebook

WRAP IT UP!

1. **Predict** Did your results support your predictions? Explain.

2. **Explain** What caused the water vapor in the air to change state?

13

ICE SCULPTOR

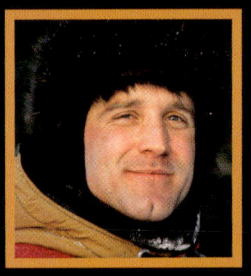

Steve Brice

Some artists make sculptures out of ice! Steve Brice carves beautiful shapes out of solid ice. He works in Chena Hot Springs in Alaska. The temperature is below freezing much of the year there. His sculptures last for months outside without melting.

Ice is hard like stone and wood. Brice uses chain saws, sanders, knives, and drills to shape the ice.

As long as it stays cold, the sculptures keep their shape. Every year, the sculptures melt in the spring. Brice makes new ones when it freezes again in the winter.

Steve Brice uses a special tool to smooth the surface of one of his ice sculptures.

Colorful lights brighten up the sculptures in Steve Brice's gallery.

Riding in a cable car is a way to experience force, motion, and a great view. Skiers and sightseers can relax in suspended cable cars as they zoom up the side of this slope in Switzerland. Motors do all the work to move the cables and the cable cars.

FORCE, MOTION, AND ENERGY

REPORTING CATEGORY 2: FORCE, MOTION, AND ENERGY

The student will demonstrate an understanding of force, motion, and energy and their relationships.

3.6 FORCE, MOTION, and ENERGY
The student knows that forces cause change and that energy exists in many forms.

WORK: SWINGS

Look at the photo of the man pushing his daughter on the swing. Is any **work** being done? Yes! In science, work is done when a force pushes or pulls an object over a distance. The man pushing his daughter on the swing is definitely doing work!

The man pushing his daughter on the swing is applying force to make her and the swing move.

voCAB

work
(WERK)

Work is done when a force is used to move an object over a distance.

SUPPORTING STANDARD TEKS 3.6.B:
Demonstrate and observe how position and motion can be changed by pushing and pulling objects to show work being done such as swings, balls, pulleys, and wagons.

The man is applying force to the girl on the swing. The girl and the swing move, or change position. The man is doing work when he pushes his daughter on the swing. Anytime that a force is used to move an object over a distance, work is being done.

My science notebook

WRAP IT UP!

1. **Define** What is work?

2. **Recall** How is pushing a person on a swing an example of work being done?

3. **Apply** Suppose you go to the grocery store. To go into the store, you push open the door. How is pushing open the door an example of work being done?

19

INVESTIGATE
MOTION OF A
PENDULUM

 **How can you show that work can be done with a pendulum?**

A pendulum is like a swing. A pendulum has a weight that is attached to the end of a string, rod, or other object. The pendulum can move back and forth like a playground swing. Work can be done by swings and pendulums. Remember, work is being done when a force is used to move or change the position of an object.

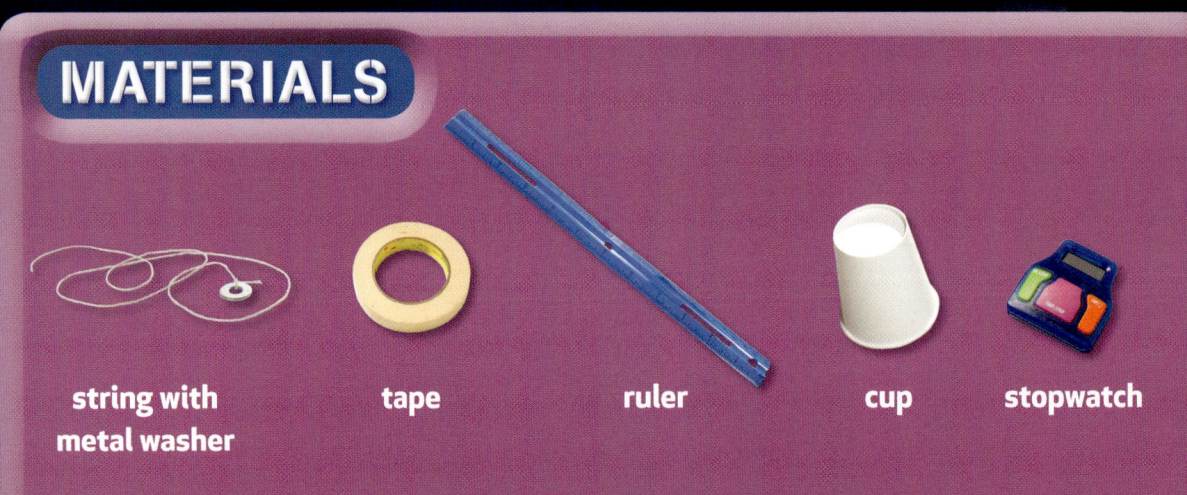

MATERIALS

string with metal washer   **tape**   **ruler**   **cup**   **stopwatch**

SUPPORTING STANDARD TEKS 3.6.B: Demonstrate and observe how position and motion can be changed by pushing and pulling objects to show work being done such as swings, balls, pulleys, and wagons.

1 Tape the free end of a string with a washer to the edge of a table. The metal washer should be close to the floor but not touching it. You have made a pendulum.

2  Pull the pendulum back 30 cm and then let it go. Observe the motion of the pendulum for one minute. Record your observations in your science notebook.

3 Place a paper cup on the floor under the edge of the table. Pull the pendulum back 30 cm.

4 Let the pendulum go. Observe the motion of the pendulum and the cup. Record your observations.

WRAP IT UP!

1. **Explain**   What force causes the pendulum to move?

2. **Explain**   How can you tell work is done as the pendulum swings? How can you tell work is done as the pendulum strikes the cup?

WORK:
WAGONS AND BALLS

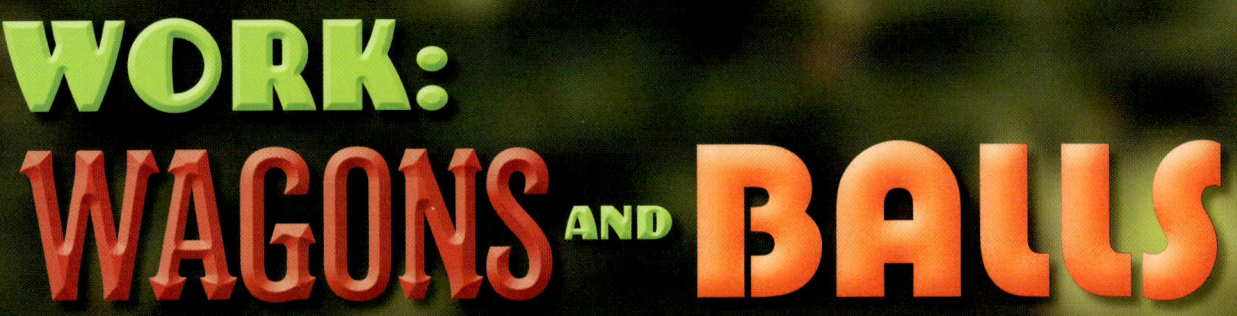

Look at the children in the photograph. They are pulling the dog in a wagon. The children are using a force to make the wagon move, or change position. Since the children are using a force to make the wagon move, the children are doing work.

SUPPORTING STANDARD TEKS 3.6.B:
Demonstrate and observe how position and
motion can be changed by pushing and pulling
objects to show work being done such as swings,
balls, pulleys, and wagons.

Work can be done with balls, too. Look at the photo of the men playing basketball. The man who is dribbling the basketball pushed the ball toward the ground. The ball moved. The man used a force to move the basketball. The man did work.

The man dribbling the basketball did work because he used a force to push the ball down.

 **WRAP IT UP!**

1. **Recall**   How is pulling a wagon an example of work being done?

2. **Explain**   How can kicking a football be work?

3. **Apply**   Think of an activity that you do at school. It might be opening your notebook or playing a drum. Explain whether that activity is work, and why.

WORK: PULLEYS

These window washers use **pulleys** to move up and down the building. A pulley is a grooved wheel with a cable or a rope running through the groove. Since the window washers are moving themselves up the building using the pulleys, the window washers are doing work with the pulleys.

Pulleys control the height of the window washers on an office building.

voCAB

pulley
(PUL-lē)

A **pulley** is a grooved wheel with a cable or a rope running through the groove.

READY SET STAAR

SUPPORTING STANDARD TEKS 3.6.B:
Demonstrate and observe how position and motion can be changed by pushing and pulling objects to show work being done such as swings, balls, pulleys, and wagons.

When the window washers want to move up the building, they pull on the ropes. Since the ropes are attached to pulleys, it makes the work of moving up the building easier. The window washers pull down on the ropes, and they move up the building.

Pulleys can be used to do work to move objects more easily.

My science notebook

WRAP IT UP !

1. **Recall** How is using a pulley to lift an object an example of work being done?

2. **Explain** People can use pulleys to close the curtains at a theater. How is this an example of work being done?

3. **Apply** If you try to lift an object using a pulley but the object does not move, has work been done? Explain.

25

ROLLER COASTER DESIGNER

Cynthia Emerick

As pushes and pulls move roller coasters, work is being done. But it is fun work! Roller coaster designer Cynthia Emerick takes advantage of different forces to design roller coasters that are thrilling to ride.

NG Science: What is your job?

Cynthia Emerick: I oversee the design and installation of roller coasters. I figure out how to do it in a way that's safe and not too expensive. My team and I solve engineering problems every day.

NG Science: What did you do in school to learn how to do your job?

Cynthia Emerick: I always found math and science interesting. In college, I learned about materials and how to put them together to make things. I also learned about what happens when materials break down. That helps me understand how to make rides safe.

This roller coaster has powerful motors that move the cars to the top of the first hill. From there, gravity takes over.

Work is done as forces move roller coasters in all directions.

27

As lava from the volcano Kilauea, located on the island of Hawai'i, reaches the ocean, an explosion of steam sprays pieces of lava into the air. The erupting lava forms new land as it flows onto the surrounding areas and into the ocean.

EARTH AND SPACE

REPORTING CATEGORY 3: EARTH AND SPACE

The student will demonstrate an understanding of components, cycles, patterns, and natural events of Earth and space systems.

3.7 EARTH and SPACE
The student knows that Earth consists of natural resources and its surface is constantly changing.

3.8 EARTH and SPACE
The student knows that there are recognizable patterns in the natural world and among objects in the sky.

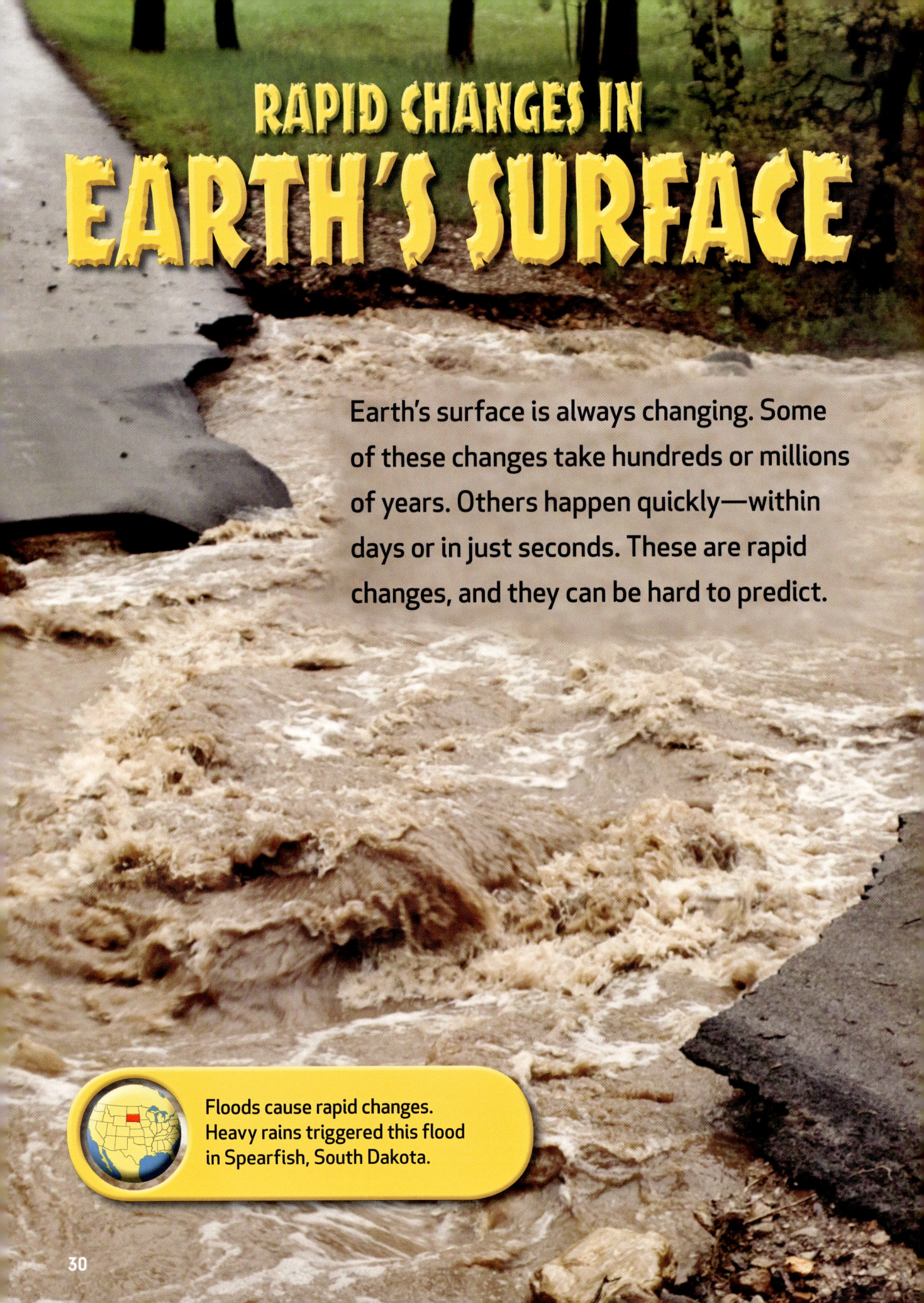

RAPID CHANGES IN
EARTH'S SURFACE

Earth's surface is always changing. Some of these changes take hundreds or millions of years. Others happen quickly—within days or in just seconds. These are rapid changes, and they can be hard to predict.

Floods cause rapid changes. Heavy rains triggered this flood in Spearfish, South Dakota.

SUPPORTING STANDARD TEKS 3.7.B: Investigate rapid changes in Earth's surface such as volcanic eruptions, earthquakes, and landslides.

READY SET STAAR

Some rapid changes are caused by weather, such as floods from heavy rain. Floods can move large amounts of soil and rock. Other rapid changes are caused by what happens far below Earth's surface.

My science notebook

WRAP IT UP !

1. **Recall** How quickly do some rapid changes to Earth's surface happen?

2. **Explain** How can heavy rain cause rapid changes in Earth's surface?

3. **Conclude** How has Earth's surface changed in this picture?

VOLCANOES!

A volcano is an opening in Earth's surface through which materials can erupt. What is an **eruption?** A volcano erupts when melted rock, ash, and gases move up through the volcano onto the surface of Earth. Eruptions cause rapid changes to Earth's surface.

Mt. Mayon is the most active volcano in the Philippines.

eruption
(i-RUP-shun)

An **eruption** is the movement of melted rock, ash, and gases up through a volcano.

magma
(MAG-mah)

Magma is melted rock beneath Earth's surface.

lava
(LAH-vah)

Lava is melted rock that flows from a volcano onto Earth's surface.

magma

READY SET STAAR

SUPPORTING STANDARD TEKS 3.7.B: Investigate rapid changes in Earth's surface such as volcanic eruptions, earthquakes, and landslides.

Look at the diagram below. **Magma** is melted rock beneath Earth's surface. The magma that erupts onto the surface is called **lava.** As lava cools, it hardens into rock, which builds up the surface of the land.

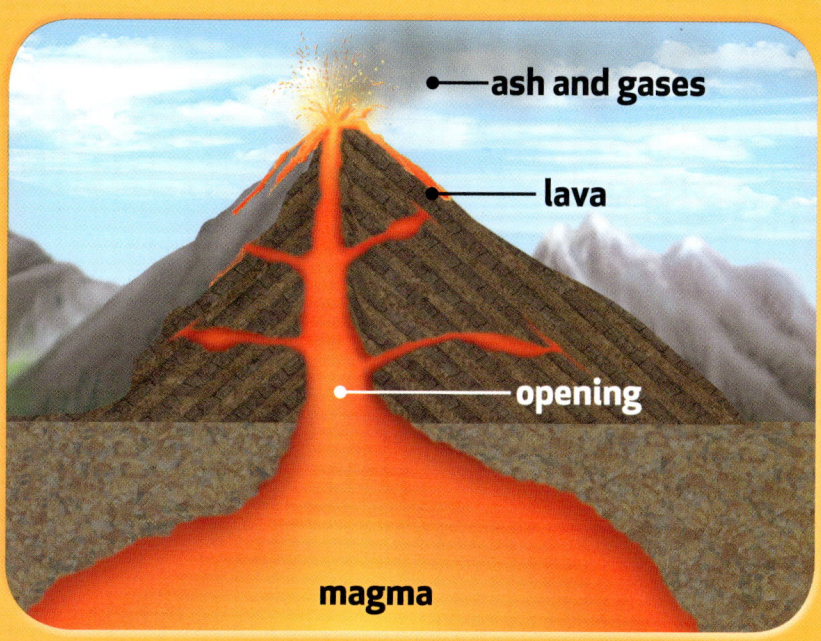

Magma rises up a volcano and erupts as lava. Ash and gases erupt into the air.

my science notebook **WRAP IT UP!**

1. **Recall**   What kinds of materials erupt from volcanoes?

2. **Compare and Contrast**   How are magma and lava similar? How are they different?

3. **Infer**   Lava hardens into new rock. If a volcano erupts many times, how might the size of the volcano change over time?

VOLCANOES

CHANGE EARTH'S SURFACE

Eruptions are rapid changes. When volcanoes erupt, they can change Earth's surface quickly over a few days, or in just minutes. In a powerful eruption, lava and ash can cover and change the land.

before

This valley on Montserrat was lush and green before the Soufriere Hills volcano erupted.

after

Here is the same valley after the eruption. The land is blanketed with ash.

READY SET STAAR

SUPPORTING STANDARD TEKS 3.7.B:
Investigate rapid changes in Earth's surface such as volcanic eruptions, earthquakes, and landslides.

Rivers of lava can pour out of erupting volcanoes and onto Earth's surface. The lava burns everything in its path. When the lava cools, it hardens into new rock. Ash can often cover huge areas of Earth's surface when volcanoes erupt, killing plants and animals and destroying buildings. Ash can also enrich the soil, making it better for growing new plants.

Montserrat is an island in the Caribbean. The Soufriere Hills volcano is on Montserrat. It has been erupting on and off since 1995.

My science notebook

WRAP IT UP !

1. **Explain**   How do eruptions of volcanoes cause rapid changes in Earth's surface?

2. **Infer**   The Soufriere Hills volcano has erupted many times in its past. Compare the before-and-after photos of just one eruption. Infer whether or not plants and animals will return.

EARTHQUAKES!

Think about what it would be like if Earth's surface began to shake and move. That is what can happen during an **earthquake.** An earthquake can cause rapid changes in Earth's surface. It is the shaking of Earth's surface caused by sudden movement of rock beneath the surface.

This is the Greendale fault in New Zealand. An earthquake started along this fault.

earthquake
(ERTH-kwāk)

An **earthquake** is the shaking of Earth's surface caused by sudden movement of rock beneath the surface.

fault
(FAWLT)

A **fault** is a break in Earth's surface along which very large slabs of rock can move.

READY SET STAAR

SUPPORTING STANDARD TEKS 3.7.B: Investigate rapid changes in Earth's surface such as volcanic eruptions, earthquakes, and landslides.

Earthquakes start along a **fault.** A fault is a break in Earth's surface along which very large slabs of rock can move. Look at the picture. The cracks show where slabs of rock along a fault moved during an earthquake.

Science in a Snap!

Pressure Buildup

Press your thumb and middle finger together.

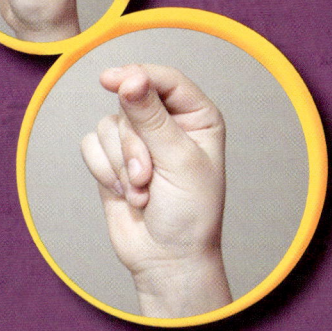

Keep pressing as you try to slide your thumb and finger past each other.

What happened as you kept pressing? How is this similar to what happens during an earthquake?

My science notebook

WRAP IT UP !

1. **Define**   In Earth science, what is a fault?

2. **Cause and Effect**   What changes beneath Earth's surface cause earthquakes?

37

EARTHQUAKES
CHANGE EARTH'S SURFACE

As huge slabs of rock move beneath Earth's surface, they can become locked together. An earthquake happens when the slabs of rock break free. This movement causes the ground to shake. Most earthquakes are too small to feel. However, some are very powerful.

An earthquake damaged this building in Christchurch, New Zealand. The earthquake started along the Greendale fault.

READY SET STAAR

SUPPORTING STANDARD TEKS 3.7.B:
Investigate rapid changes in Earth's surface such as volcanic eruptions, earthquakes, and landslides.

Earthquakes can change Earth's surface rapidly. Powerful earthquakes may crack the land or cause a landslide. A major earthquake can damage buildings and roads.

Bent train tracks show how Earth's surface changed along the Greendale fault when the earthquake happened.

My science notebook

WRAP IT UP!

1. **Explain** How can earthquakes cause rapid changes in Earth's surface?

2. **Infer** Why might earthquakes be more dangerous to people in cities than to living things in the wild?

39

LANDSLIDES

CHANGE EARTH'S SURFACE

A **landslide** is the movement of rock and soil down a slope. A landslide is a rapid change. Like earthquakes, landslides are dangerous and hard to predict. A large landslide can crush everything in its path, including buildings and other structures.

This landslide in El Salvador was started by an earthquake.

VOCAB

landslide
(LAND-slīd)

A **landslide** is the movement of rock and soil down a slope.

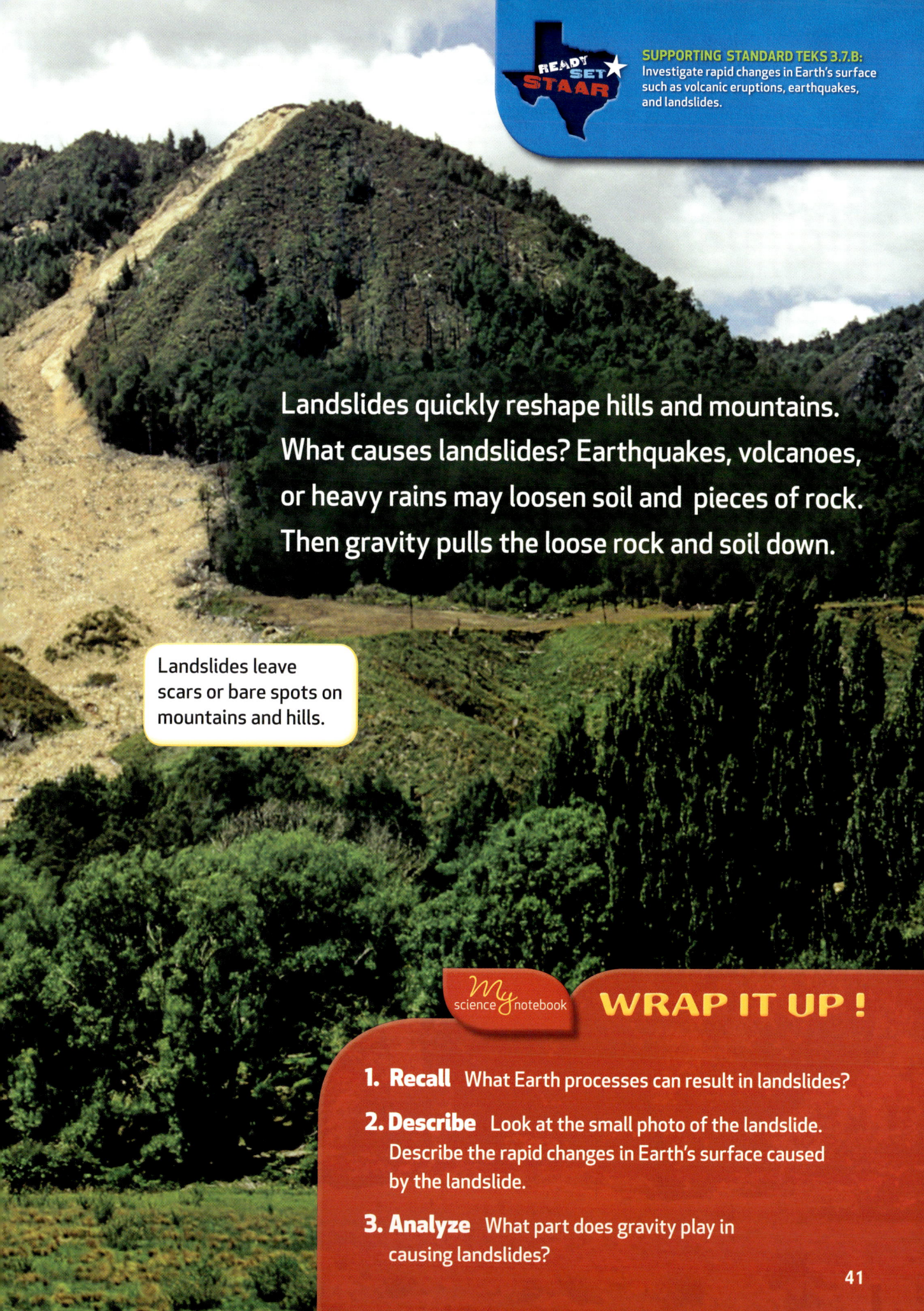

SUPPORTING STANDARD TEKS 3.7.B: Investigate rapid changes in Earth's surface such as volcanic eruptions, earthquakes, and landslides.

READY SET STAAR

Landslides quickly reshape hills and mountains. What causes landslides? Earthquakes, volcanoes, or heavy rains may loosen soil and pieces of rock. Then gravity pulls the loose rock and soil down.

Landslides leave scars or bare spots on mountains and hills.

My science notebook   WRAP IT UP !

1. **Recall**   What Earth processes can result in landslides?

2. **Describe**   Look at the small photo of the landslide. Describe the rapid changes in Earth's surface caused by the landslide.

3. **Analyze**   What part does gravity play in causing landslides?

INVESTIGATE
LANDSLIDES

? How can you model the way earthquakes affect landslides?

Earthquakes can start landslides. In a landslide, Earth's surface may change a lot or a little. The amount of change depends on the size of the landslide. In this investigation, you can use a model to learn how earthquakes can cause landslides.

MATERIALS

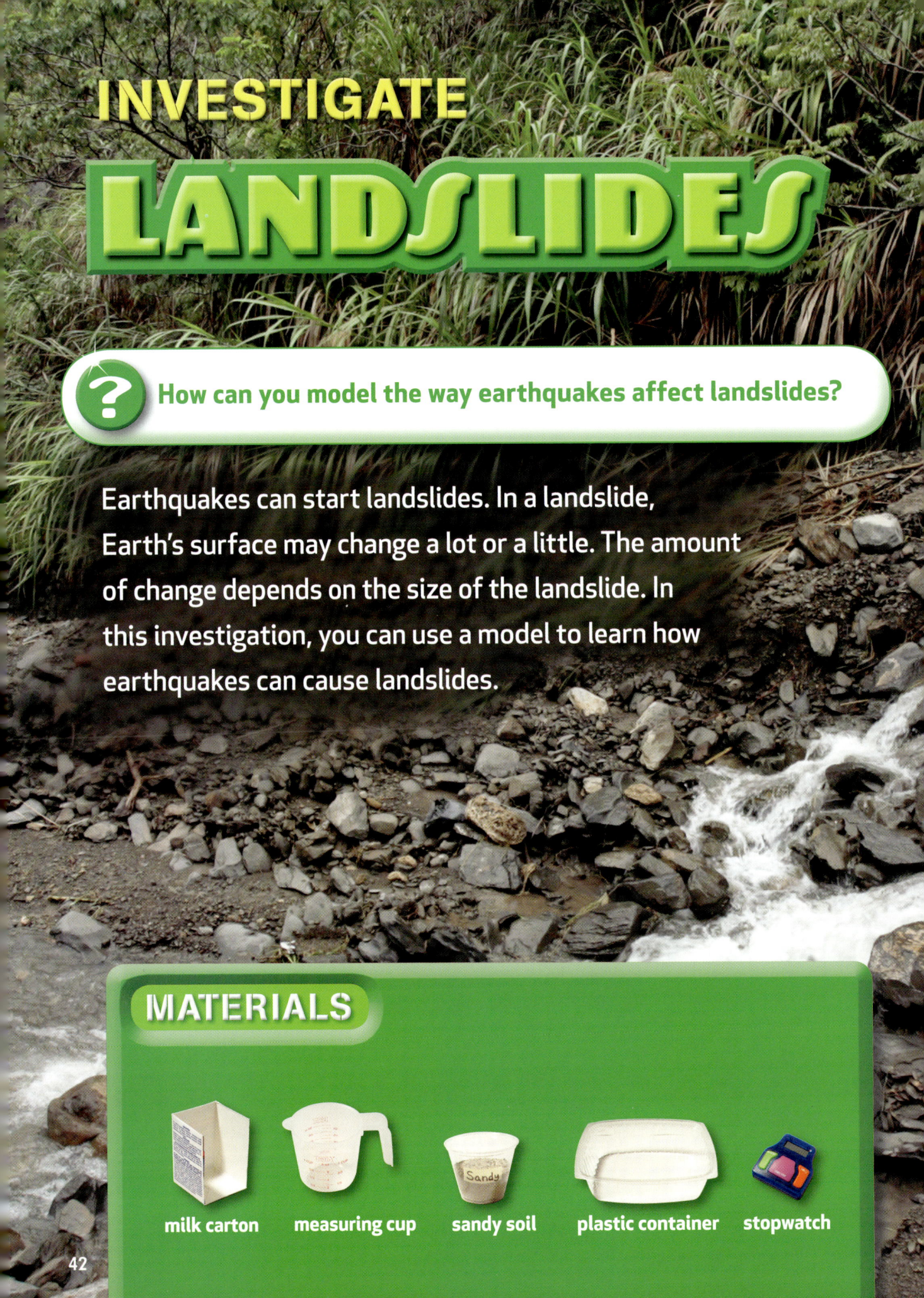

milk carton   measuring cup   sandy soil   plastic container   stopwatch

READY SET STAAR

SUPPORTING STANDARD TEKS 3.7.B: Investigate rapid changes in Earth's surface such as volcanic eruptions, earthquakes, and landslides.

1 Use a measuring cup to measure 225 mL of sandy soil. Pour the sandy soil in the closed end of a milk carton to make a hill.

2 Put the carton near the edge of a table. Have a partner hold the plastic container under the table to catch the soil.

3 My science notebook

Model a small earthquake by shaking the table gently for 30 seconds. Use a timer. Record your observations of a small earthquake in your science notebook.

4 Now model a powerful earthquake by shaking the table harder for 30 seconds. Record your observations.

My science notebook

WRAP IT UP!

1. **Infer** Use the results of this investigation to infer how Earth's surface can change during a landslide.

2. **Apply** Heavy rain is another cause of landslides. Explain how you could change the investigation to test how heavy rains change Earth's surface.

43

SOLAR SYSTEM

The solar system includes the Sun and everything that **revolves,** or moves around, the Sun. Earth is one of eight planets in the solar system. The planets in order of their distance from the Sun are Mercury, Venus, Earth, Mars, Jupiter, Saturn, Uranus, and Neptune.

Mercury

Mars

Saturn

Venus

Earth

Jupiter

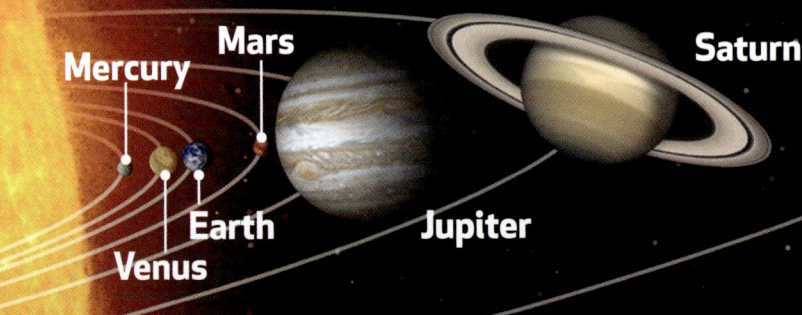

voCAB

revolve
(re-VAWLV)

To **revolve** is to travel around another object.

orbit
(ŌR-bit)

An **orbit** is the path Earth or another object takes as it revolves.

SUPPORTING STANDARD TEKS 3.8.D:
Identify the planets in Earth's solar system and
their position in relation to the Sun.

READY
SET
STAAR

Each planet revolves around the Sun in a path
called an **orbit.** You can see the orbits of the
planets in the picture.

The planets are much farther
apart than shown here. The
more distant a planet is from
the Sun, the longer its orbit.

Neptune

Uranus

WRAP IT UP !

1. Identify Name the planets in Earth's solar system and
their position in relation to the Sun.

2. Apply Which planets have smaller orbits than Earth?

3. Infer Which planet takes the longest time to travel
around the Sun?

INNER PLANETS

The eight planets can be divided into two groups—inner planets and outer planets. The inner planets are shown in the pictures in their order from the Sun.

MERCURY

Mercury is the nearest planet to the Sun. It is very hot during the day and very cold at night.

VENUS

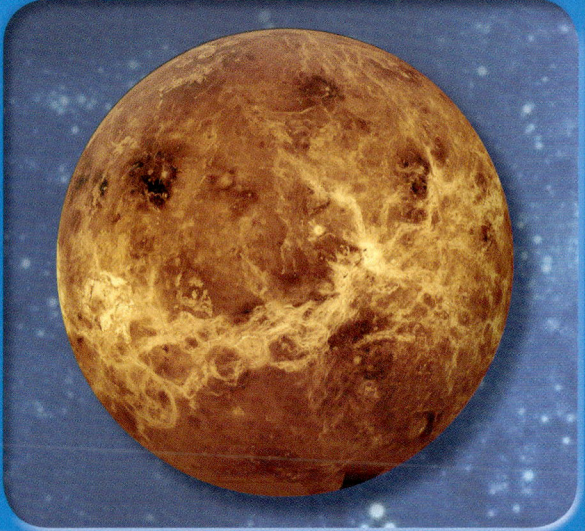

Venus is the second planet from the Sun. It is one of the brightest objects in Earth's night sky.

CLOSEST NEIGHBOR

Venus is the closest planet to Earth. Venus has thick swirling clouds that are burning hot and poisonous. It also has lightning and strong winds.

READY SET STAAR

SUPPORTING STANDARD TEKS 3.8.D:
Identify the planets in Earth's solar system and their position in relation to the Sun.

The inner planets are closer to the Sun than the outer planets. They are made mostly of rock. For this reason, they are also called the rocky planets.

EARTH

Earth is the third planet from the Sun. It is the only planet in the solar system known to support life.

MARS

The fourth planet from the Sun is Mars. It is rocky and has volcanoes, valleys, and large dust storms.

My science notebook WRAP IT UP!

1. **Identify**   Which inner planet is closest to the Sun? Which one is farthest from the Sun?

2. **Compare**   What do the inner planets have in common?

OUTER PLANETS

The outer planets are much farther from the Sun than the inner planets. The outer planets are farther apart from each other, too. They are shown in the pictures in their order from the Sun.

JUPITER

Jupiter is the fifth and biggest planet. It has a huge storm system, the Great Red Spot, that has been raging for centuries.

SATURN

The sixth planet is Saturn. Saturn has many rings. The rings are probably made of ice, dust, and rocks.

THE BIGGEST GIANT

Jupiter is gigantic. Jupiter is so large that 1,324 Earths could fit inside. Even Jupiter's Great Red Spot is bigger than Earth.

SUPPORTING STANDARD TEKS 3.8.D:
Identify the planets in Earth's solar system and their position in relation to the Sun.

READY SET STAAR

The outer planets are made mostly of gases that surround a rocky core. The outer planets are called the gas giants. They are the largest planets in the solar system.

URANUS

The seventh planet from the Sun is Uranus. Like Saturn, it has rings, but they are dark and hard to see with telescopes on Earth.

NEPTUNE

The eighth planet in the solar system is Neptune. Strong winds sweep gigantic storms across the planet.

My science notebook WRAP IT UP!

1. **Identify** Which outer planet is closest to the Sun? Which one is farthest from the Sun?

2. **Compare** What do the outer planets have in common?

3. **Contrast** How do the outer planets differ from the inner planets?

VOLCANOLOGIST

Tamsin Mather

Would you like to climb a volcano or study what's inside? Do you want to help save lives by learning to predict eruptions? Then you might want to be a volcanologist like Tamsin Mather.

NG Science: What does a volcanologist do?

Tamsin Mather: I study volcanoes and their effects on our planet. Some of my time is spent visiting volcanoes to take measurements and collect samples. I take the samples to a lab for analysis. I also teach students about volcanoes at a university.

NG Science: Do you see a strong connection between what you do and Earth science?

Tamsin Mather: Of course! Volcanoes have played a key part in the history of our planet. Major eruptions show the importance of Earth science to people's safety.

TEKS 3.3.D:
Connect grade-level appropriate science concepts with the history of science, science careers, and contributions of scientists.

Mount Etna is a volcano on the island of Sicily in Italy.

Mount Etna

Tamsin wears a mask to protect her from gases coming from the volcano.

Organisms get what they need to survive from their environment. This ladybug on a cholla cactus eats insects that feed on the plant. The cactus makes its own food using energy from the Sun. Some desert birds weave grasses and twigs into a safe nest in the prickly spines of the cactus.

ORGANISMS AND ENVIRONMENTS

REPORTING CATEGORY 4: ORGANISMS AND ENVIRONMENTS

The student will demonstrate an understanding of the structures and functions of living organisms and their interdependence on each other and on their environment.

3.9 ORGANISMS and ENVIRONMENTS
The student knows that organisms have characteristics that help them survive and can describe patterns, cycles, systems, and relationships within the environments.

3.10 ORGANISMS and ENVIRONMENTS
The student knows that organisms undergo similar life processes and have structures that help them survive within their environments.

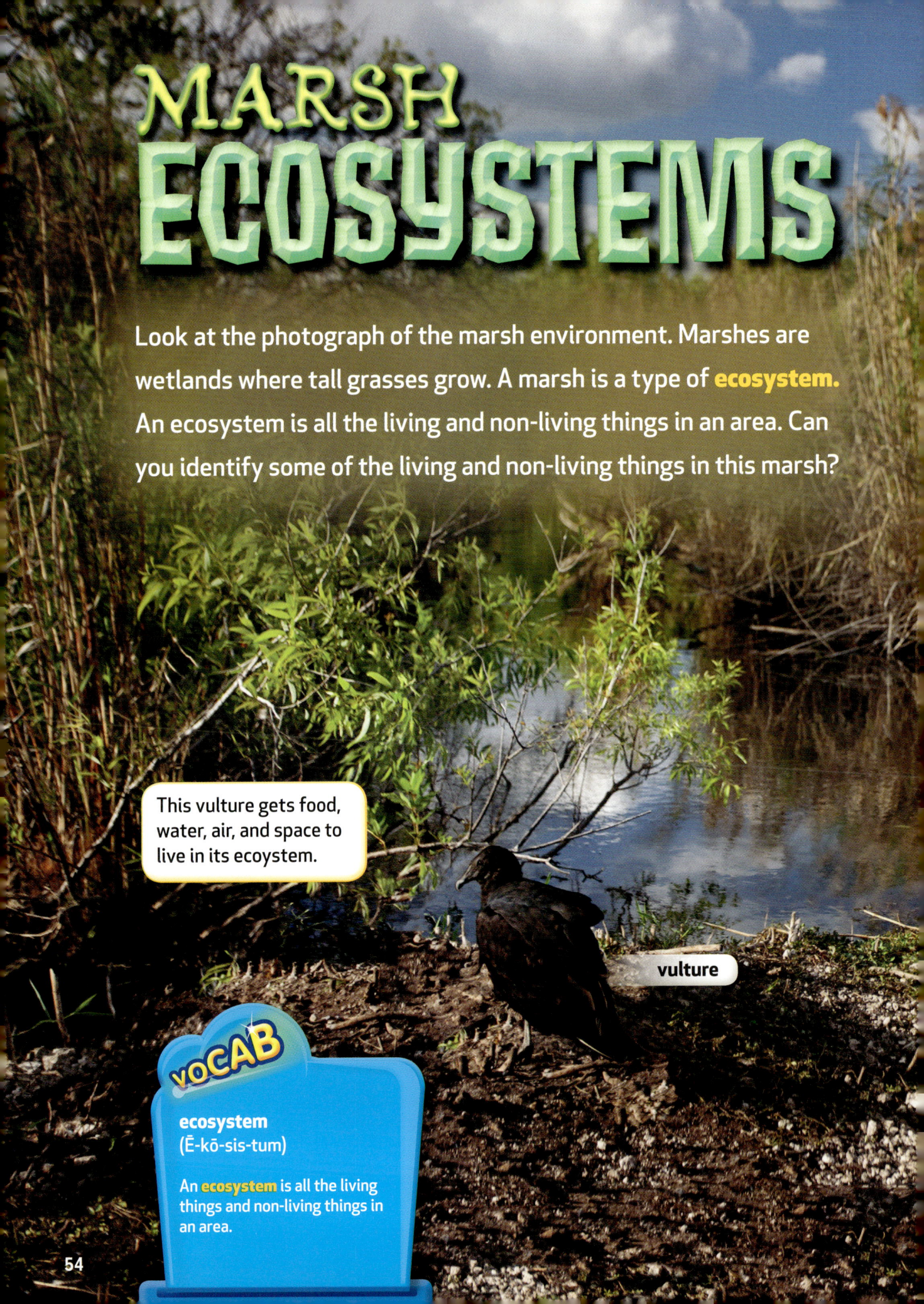

MARSH ECOSYSTEMS

Look at the photograph of the marsh environment. Marshes are wetlands where tall grasses grow. A marsh is a type of **ecosystem.** An ecosystem is all the living and non-living things in an area. Can you identify some of the living and non-living things in this marsh?

This vulture gets food, water, air, and space to live in its ecoystem.

vulture

voCAB

ecosystem
(Ē-kō-sis-tum)

An **ecosystem** is all the living things and non-living things in an area.

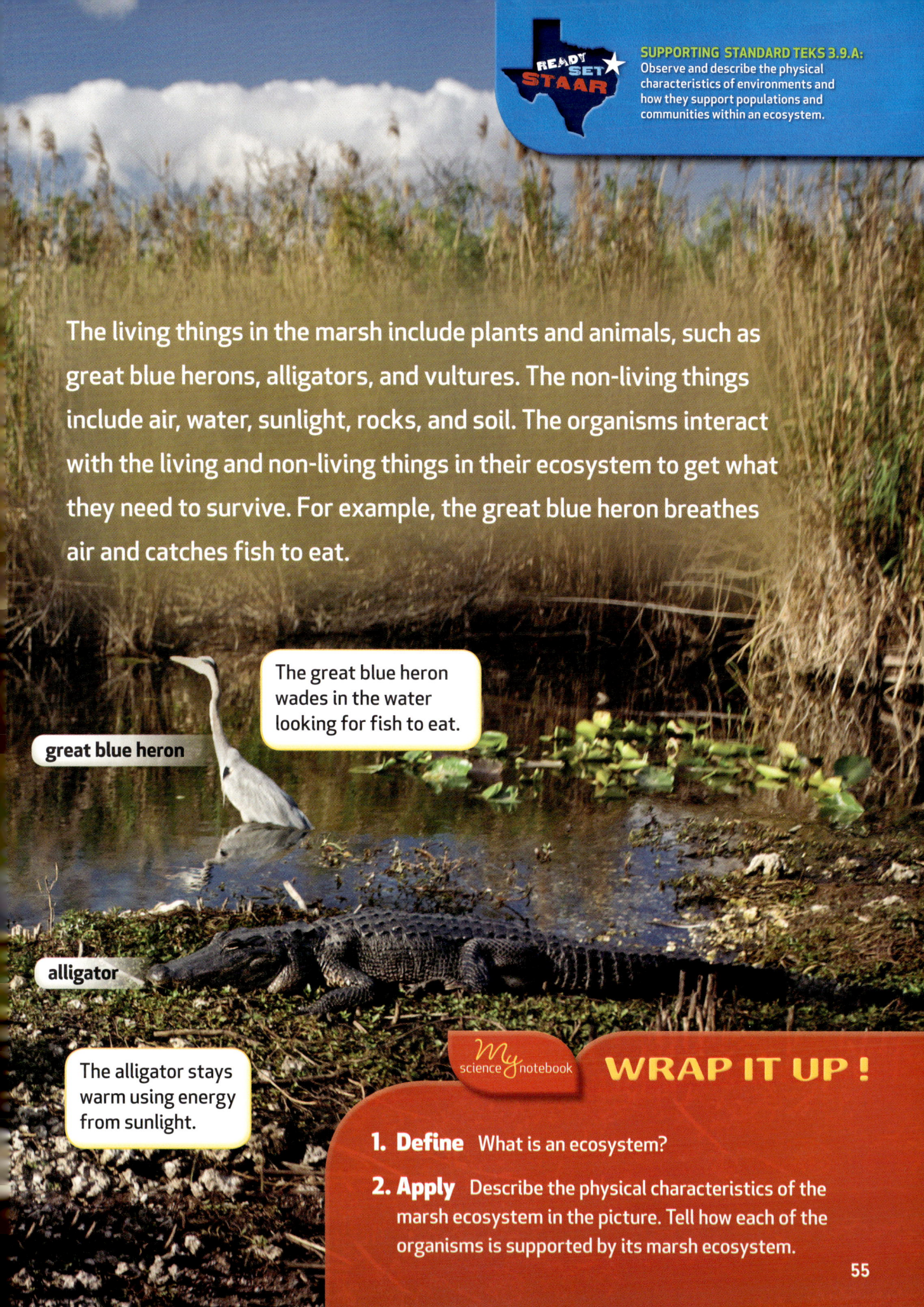

SUPPORTING STANDARD TEKS 3.9.A: Observe and describe the physical characteristics of environments and how they support populations and communities within an ecosystem.

The living things in the marsh include plants and animals, such as great blue herons, alligators, and vultures. The non-living things include air, water, sunlight, rocks, and soil. The organisms interact with the living and non-living things in their ecosystem to get what they need to survive. For example, the great blue heron breathes air and catches fish to eat.

The great blue heron wades in the water looking for fish to eat.

great blue heron

alligator

The alligator stays warm using energy from sunlight.

My science notebook

WRAP IT UP!

1. **Define** What is an ecosystem?

2. **Apply** Describe the physical characteristics of the marsh ecosystem in the picture. Tell how each of the organisms is supported by its marsh ecosystem.

GRASSLAND POPULATIONS

Many different kinds of organisms live in grasslands. Animals, such as these white-tailed deer, can live there. All of the white-tailed deer living in one area make up a **population.** A population is all the individuals of a species that live in an ecosystem. The population of white-tailed deer gets what it needs from its ecosystem. The white-tailed deer breathe the air and find water to drink.

The white-tailed deer find food, such as grass, within the ecosystem.

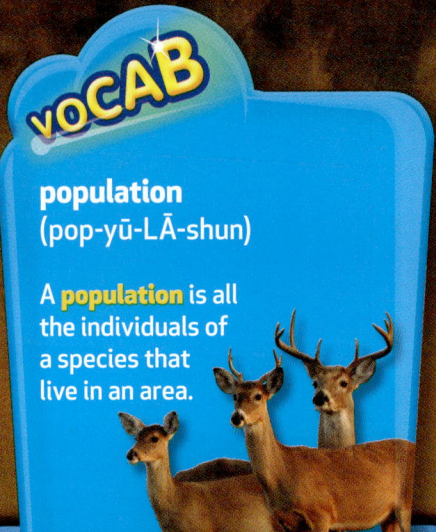

voCAB

population
(pop-yū-LĀ-shun)

A **population** is all the individuals of a species that live in an area.

SUPPORTING STANDARD TEKS 3.9.A:
Observe and describe the physical characteristics of environments and how they support populations and communities within an ecosystem.

The tall grasses and shrubs provide the white-tailed deer with a place to hide from predators.

white-tailed deer

my science notebook **WRAP IT UP!**

1. **Define**   What is a population?

2. **Describe**   How do the physical characteristics of the grassland help support the population of white-tailed deer that live there?

GRASSLAND COMMUNITY

The grassland ecosystem supports a wide variety of species, such as the whooping cranes and white-tailed deer in the photo. You have already learned that all of the white-tailed deer in an area make up a population. All of the whooping cranes in an area also make up a population.

adult whooping crane

The whooping crane and white-tailed deer populations compete for water, food, and space in their grassland ecosystem.

The population of whooping cranes and the population of white-tailed deer make up a **community.** A community is made up of all the different populations that live and interact in an area. The organisms in the grassland populations and community find the air, water, food, and space they need in their ecosystem.

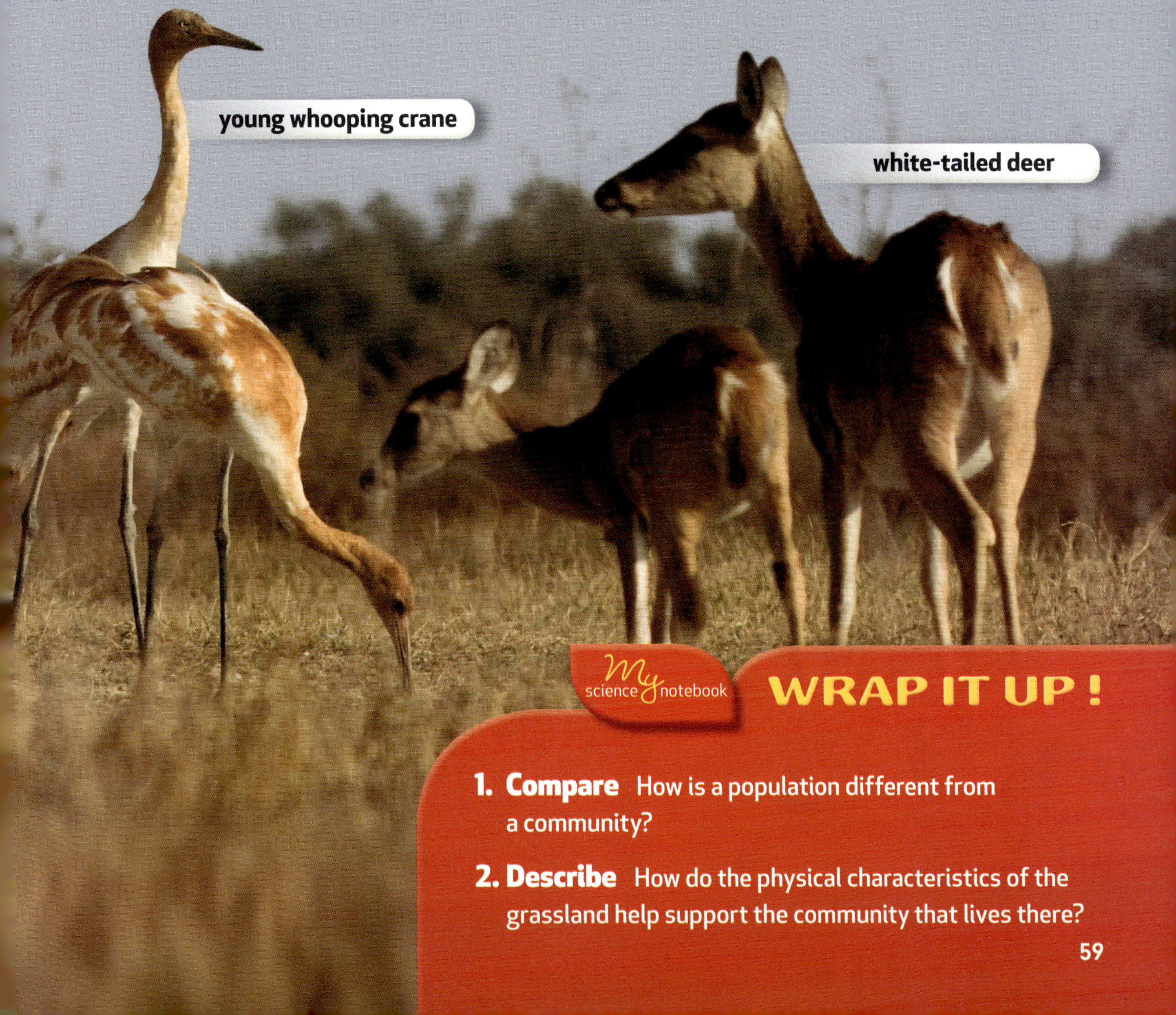

young whooping crane

white-tailed deer

WRAP IT UP!

My science notebook

1. **Compare** How is a population different from a community?

2. **Describe** How do the physical characteristics of the grassland help support the community that lives there?

59

POPULATIONS

INDIVIDUAL

POPULATION

One organism, such as this white-tailed deer, is the smallest unit of an **ecosystem.**

All the white-tailed deer that live in the grassland are a **population.**

voCAB

ecosystem (Ē-kō-sis-tum)	**population** (pop-yū-LĀ-shun)
An **ecosystem** is all the living and non-living things in an area.	A **population** is all the individuals of a species that live in an area.

COMMUNITIES

READY SET STAAR

SUPPORTING STANDARD TEKS 3.9.A: Observe and describe the physical characteristics of environments and how they support populations and communities within an ecosystem.

COMMUNITY

All the populations of living things that live and interact in the grassland form a **community.**

ECOSYSTEM

All the living and non-living things in this grassland get what they need to survive from their ecosystem.

SHARE AND COMPARE

- Choose a plant or animal.
- Research that plant or animal. Find out more about the community the living thing is part of and how the physical characteristics of its ecosystem allows the living thing to get what it needs to survive.
- Present your research about the plant or animal to the class.

community
(kuh-MYŪ-nuh-tē)

A **community** is made up of all the different populations that live and interact in an area.

61

Life Cycle
of a Tomato Plant

Many flowering plants go through similar stages of life. The series of stages that a tomato plant goes through during its lifetime is its **life cycle.** Follow the diagram of the life cycle of the tomato plant as you read.

In some parts of Texas, tomato plants can grow year-round.

SUPPORTING STANDARD TEKS 3.10.C:
Investigate and compare how animals and plants undergo a series of orderly changes in their diverse life cycles such as tomato plants, frogs, and ladybugs.

Life Cycle of a
Tomato Plant

Seed

Each tomato fruit holds many seeds.

Adult Plant

Flowers can grow on an adult tomato plant. The flowers produce fruit called tomatoes.

Seedling

A seed that is planted in soil can grow into a seedling.

Young Plant

The seedling then grows into a young plant. Its broad leaves branch out to get sunlight.

WRAP IT UP!

1. **Define**   What is a life cycle?

2. **Infer**   Do you think the life cycle of the tomato plant ends? On what evidence do you base your inference?

63

INVESTIGATE
The Life Cycle
of a Marigold

 **How does a marigold change as it grows?**

Organisms undergo many changes during their lifetime. These changes in an organism's life are called its life cycle. In this investigation, you will observe the changes that occur during the life cycle of a marigold.

MATERIALS

marigold seeds

hand lens

cup with soil

spoon

SUPPORTING STANDARD TEKS 3.10.C:
Investigate and compare how animals and plants undergo a series of orderly changes in their diverse life cycles such as tomato plants, frogs, and ladybugs.

1

my science notebook

Observe marigold seeds with a hand lens. Record your observations in your science notebook. Use a spoon to plant the seeds in a cup with soil.

2

Place the cup in a sunny place. Keep the soil moist. When the seeds sprout, observe and draw the plant parts.

3

As the plants grow, observe how they change. Make drawings of your observations.

4

When the marigolds bloom, use a hand lens to observe the parts of the flowers.

my science notebook

WRAP IT UP!

1. **Explain**   What changes did you observe in the plants as they grew?

2. **Compare**   How are the life cycles of the marigold and tomato plant alike? How are they different?

Life Cycle of a Ladybug

Ladybugs are small, spotted, oval-shaped insects. The ladybug looks different during each stage of its life cycle. Trace the diagram of the ladybug life cycle as you read about each stage.

Ladybugs eat many different types of insects, such as aphids and mealybugs, that attack crops.

SUPPORTING STANDARD TEKS 3.10.C:
Investigate and compare how animals and plants undergo a series of orderly changes in their diverse life cycles such as tomato plants, frogs, and ladybugs.

Adult

An adult ladybug has wings and can fly. It looks very different from the other stages.

Egg

An adult female ladybug lays its eggs on a leaf.

Life Cycle of a
Ladybug

Larva

The ladybug **larva** may eat small insects. It sheds its outer covering as it grows.

Pupa

The ladybug changes form during the **pupa** stage.

My science notebook

WRAP IT UP !

1. **Define**   What is a larva?

2. **Contrast**   Describe some differences between the pupa and the adult stages in the ladybug life cycle.

LIFE CYCLE OF A LEOPARD FROG

Frogs and other amphibians go through a life cycle in which the animal looks different at different stages. Look at the photos of the life cycle of the leopard frog. Trace the diagram of its life cycle with your finger as you read about each stage.

Leopard frogs use their strong hind legs to escape danger.

READY SET STAR

SUPPORTING STANDARD TEKS 3.10.C:
Investigate and compare how animals and plants undergo a series of orderly changes in their diverse life cycles such as tomato plants, frogs, and ladybugs.

Egg
An adult female leopard frog lays its eggs in a pond or swamp.

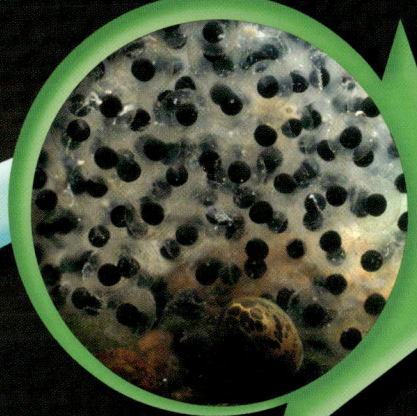

Adult
An adult leopard frog lives on land and breathes air.

LIFE CYCLE OF A LEOPARD FROG

Tadpole
A tadpole has a tail and no legs. It lives under water and breathes through gills.

Young Frog
The young leopard frog begins to grow legs. Its tail begins to shorten.

My science notebook

WRAP IT UP!

1. **Recall**   What are the stages in the life cycle of a frog?

2. **Contrast**   Describe some differences between the tadpole stage and the adult stage of the frog.

ZOO CURATOR

Maria Franke

Maria Franke is the Curator of Mammals at the Toronto Zoo in Canada. She helps design and build new environments for the animals in the zoo. She and her team also raise young black-footed ferrets, which are endangered due to loss of habitat. The young ferrets are released into the wild.

NG Science: What do you do as the Curator of Mammals at the Toronto Zoo?

Maria Franke: Every day is different. I make sure we have healthy populations of different animals. I help design and plan exhibits. I work with keepers to build habitats that improve the lives of the animals. I give talks about the zoo's conservation work. I also work to save endangered animals.

NG Science: What part of your job is most important to you?

Maria Franke: The coolest part is helping to save endangered species. Helping to save the black-footed ferrets in Canada was a real thrill.

Maria and another member of her team release a black-footed ferret back into the wild.

black-footed ferret

GLOSSARY

C

community (kuh-MYŪ-nuh-tē)

A community is made up of all the different populations that live and interact in an area. (pp. 58, 61)

condensation (kon-den-SĀ-shun)

Condensation is the change from a gas to a liquid. (p. 10)

E

earthquake (ERTH-kwāk)

An earthquake is the shaking of Earth's surface caused by sudden movement of rock beneath the surface. (p. 36)

ecosystem (Ē-kō-sis-tum)

An ecosystem is all the living things and non-living things in an area. (pp. 54, 60)

eruption (i-RUP-shun)

An eruption is the movement of melted rock, ash, and gases up through a volcano onto Earth's surface. (p. 32)

evaporation (i-vap-uh-RĀ-shun)

Evaporation is the change from a liquid to a gas. (p. 6)

F

fault (FAWLT)

A fault is a break in Earth's surface along which very large slabs of rock can move. (p. 36)

L

landslide (LAND-slīd)

A landslide is the movement of rock and soil down a slope. (p. 40)

larva (LAR-vah)

A larva is a young animal with a body form very different from the adult. (p. 66)

lava (LAH-vah)

Lava is melted rock that flows from a volcano onto Earth's surface. (p. 32)

life cycle (LĪF SĪ-kul)

A life cycle is a series of stages through which an organism passes during its lifetime. (p. 62)

M

magma (MAG-mah)

Magma is melted rock beneath Earth's surface. (p. 32)

matter (MA-ter)

Matter is anything that has mass and takes up space. (p. 4)

O

orbit (ŌR-bit)

An orbit is the path Earth or another object takes as it revolves. (p. 44)

P

population (pop-yū-LĀ-shun)

A population is all the individuals of a species that live in an area. (pp. 56, 60)

pulley (PUL-lē)

A pulley is a grooved wheel with a cable or a rope running through the groove. (p. 24)

pupa (PYŪ-pah)

A pupa is the stage of a life cycle in which the body of a young animal changes from larva to adult. (p. 66)

R

revolve (re-VAWLV)

To revolve is to travel around another object. (p. 44)

S

states of matter (STĀTS UV MA-ter)

States of matter are the forms in which a material can exist. (p. 4)

W

work (WERK)

Work is done when a force is used to move an object over a distance. (p. 18)

INDEX

Photographs

PROGRAM CONSULTANTS

Randy Bell, Ph.D.
Associate Professor of Science Education,
University of Virginia, Charlottesville, Virginia
SCIENCE

Kathy Cabe Trundle, Ph.D.
Associate Professor of Early Childhood
Science Education,
The School of Teaching and Learning,
The Ohio State University, Columbus, Ohio
SCIENCE

Judith Sweeney Lederman, Ph.D.
Director of Teacher Education,
Associate Professor of Science Education,
Department of Mathematics and
Science Education,
Illinois Institute of Technology, Chicago, Illinois
SCIENCE

David W. Moore, Ph.D.
Professor of Education,
Mary Lou Fulton Teachers College,
Arizona State University, Tempe, Arizona
LITERACY

PROGRAM CONTRIBUTOR

Cathey Whitener, M.S. in Ed.
Science Specialist,
Marcella Intermediate School,
Aldine ISD, Houston, Texas
SCIENCE

Acknowledgments
Grateful acknowledgment is given to the authors, artists, photographers, museums, publishers, and agents for permission to reprint copyrighted material. Every effort has been made to secure the appropriate permission. If any omissions have been made or if corrections are required, please contact the Publisher.

STAAR is a trademark and/or federally registered trademark owned by the Texas Education Agency, and is used pursuant to license.

Photographic Credits
Front cover ©Rolf Nussbaumer/ Nature Picture Library. (bkg) ©Siede Preis/Getty Images. **Back cover** ©Rolf Nussbaumer/Nature Picture Library.

Illustrator Credit
Precision Graphics.

Maps Credit
Mapping Specialists.

Acknowledgments and credits continued on page 81.

For permission to use material from this text or product, submit all requests online at www.cengage.com/ permissions

Further permissions questions can be emailed to permissionrequest@ cengage.com

Visit National Geographic Learning online at www.NGSP.com

Visit our corporate website at www. cengage.com

Printed in the USA.
RR Donnelley, Jefferson City, MO

ISBN: 978-07362-93860

12 13 14 15 16 17 18 19 20 21

10 9 8 7 6 5 4 3 2 1